BODIES OF WATER

RIVERS

A Crabtree Roots Book

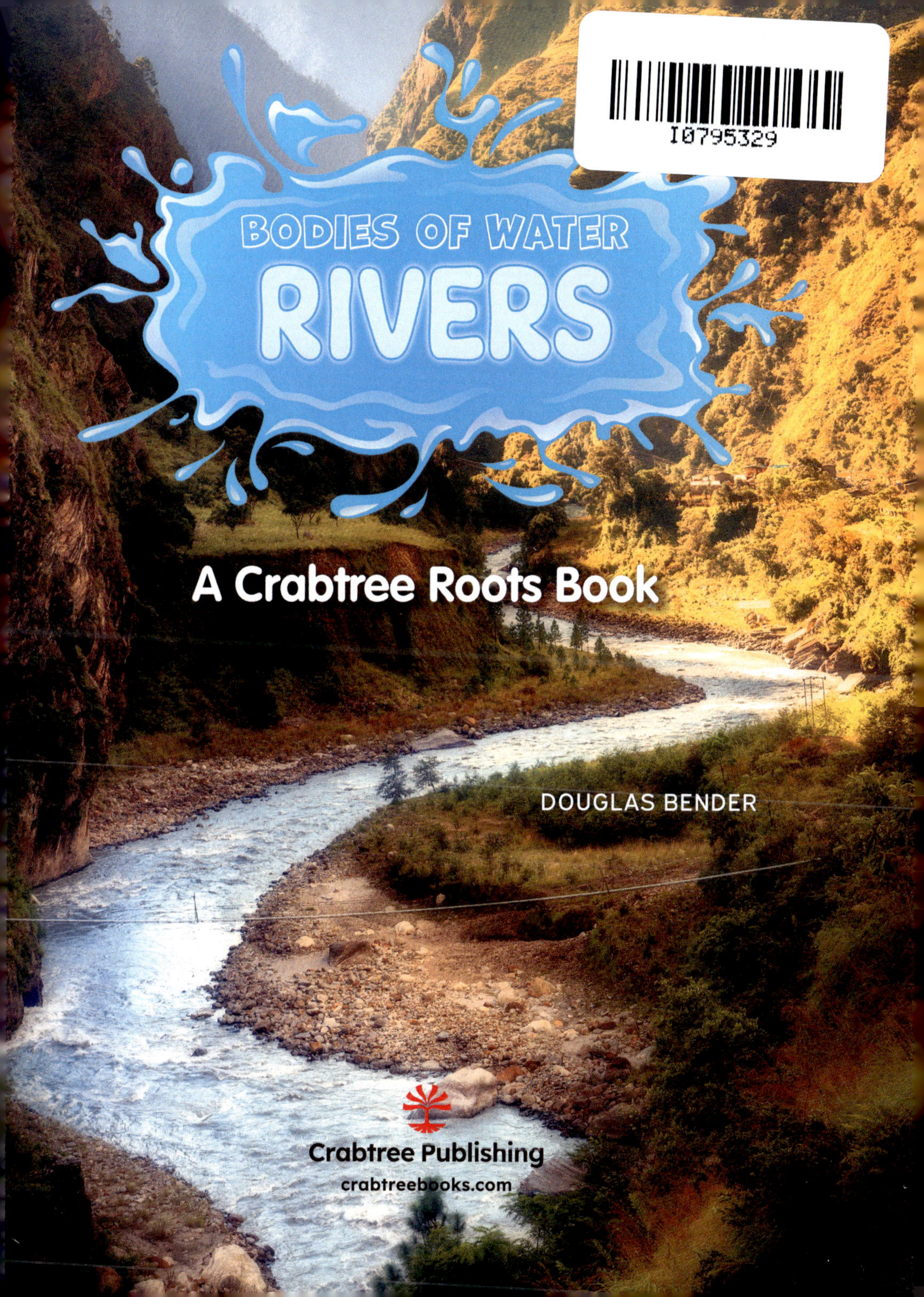

DOUGLAS BENDER

Crabtree Publishing
crabtreebooks.com

School-to-Home Support for Caregivers and Teachers

This book helps children grow by letting them practice reading. Here are a few guiding questions to help the reader with building his or her comprehension skills. Possible answers appear here in red.

Before Reading:

- What do I think this book is about?
    - *I think this book is about water and rivers.*
    - *I think this book is about what rivers look like.*
- What do I want to learn about this topic?
    - *I want to learn what animals live in rivers.*
    - *I want to learn what size a river is.*

During Reading:

- I wonder why...
    - *I wonder if rivers can be short.*
    - *I wonder how fish swim in fast-moving rivers.*
- What have I learned so far?
    - *I have learned that many rivers have waterfalls.*
    - *I have learned that rivers are long.*

After Reading:

- What details did I learn about this topic?
    - *I have learned that many rivers have rapids.*
    - *I have learned that most rivers have rocks.*
- Read the book again and look for the vocabulary words.
    - *I see the word **waterfalls** on page 7 and the word **rapids** on page 12. The other vocabulary words are found on page 14.*

This is a **river**.

Some rivers are long.

In many rivers you can find **waterfalls**.

Most rivers have **rocks**.

In some rivers you can find fish.

Many rivers
have **rapids**.

Word List

Sight Words

a	have	some
are	in	this
can	is	you
find	long	
fish	many	

Words to Know

rapids

river

rocks

waterfalls

30 Words

This is a **river**.

Some rivers are long.

In many rivers you can find **waterfalls**.

Most rivers have **rocks**.

In some rivers you can find fish.

Many rivers have **rapids**.

BODIES OF WATER

RIVERS

Crabtree Publishing

crabtreebooks.com 800-387-7650

 In Canada: We acknowledge the financial support of the Government of Canada through the Canada Book Fund for our publishing activities.

Hardcover	978-1-4271-5594-8
Paperback	978-1-4271-5600-6
Ebook (pdf)	978-1-4271-3379-3
Epub	978-1-4271-3439-4
Read-along	978-1-4271-5618-1
Audio book	978-1-4271-5624-2

Library and Archives Canada Cataloguing in Publication
Title: Rivers / Douglas Bender.
Names: Bender, Douglas, 1992- author.
Description: Series statement: Bodies of water | "A Crabtree roots book".
Identifiers: Canadiana (print) 20210190736 | Canadiana (ebook) 20210190752 | ISBN 9781427155948 (hardcover) | ISBN 9781427156006 (softcover) | ISBN 9781427133793 (HTML) | ISBN 9781427134394 (EPUB) | ISBN 9781427156181 (read-along ebook)
Subjects: LCSH: Rivers—Juvenile literature.
Classification: LCC GB1203.8 .B45 2022 | DDC j551.48/3—dc23

Published in Canada
Crabtree Publishing
616 Welland Avenue
St. Catharines, Ontario
L2M 5V6

Published in the United States
Crabtree Publishing
347 Fifth Avenue
Suite 1402-145
New York, NY 10016

Written by: Douglas Bender
Designed by: Rhea Wallace
Series Development: James Earley
Proofreader: Janine Deschenes
Educational Consultant: Marie Lemke M.Ed.
Photographs: Shutterstock: Smelov: cover; Denis Belitsky: p. 1; Liudmila Yagovating: p. 3, 14; Alexandr Vorobev: p. 5; 4045: p. 6, 14; Nikolay007: p. 8-9, 14; Menno Schaefer: p. 10, 14; BBrown: p. 13, 14

Printed in Canada/082024/CPC20240821

Library of Congress Cataloging-in-Publication Data
Names: Bender, Douglas, 1992- author.
Title: Rivers / Douglas Bender.
Description: New York, NY : Crabtree Publishing Company, [2022] | Series: Bodies of water - a crabtree roots book | Includes index.
Identifiers: LCCN 2021017130 (print) | LCCN 2021017131 (ebook) ISBN 9781427155948 (hardcover) | ISBN 9781427156006 (paperback) | ISBN 9781427133793 (ebook) | ISBN 9781427134394 (epub) | ISBN 9781427156181
Subjects: LCSH: Rivers--Juvenile literature.
Classification: LCC GB1203.8 .B45 2022 (print) | LCC GB1203.8 (ebook) | DDC 551.48/3--dc23
LC record available at https://lccn.loc.gov/2021017130
LC ebook record available at https://lccn.loc.gov/2021017131